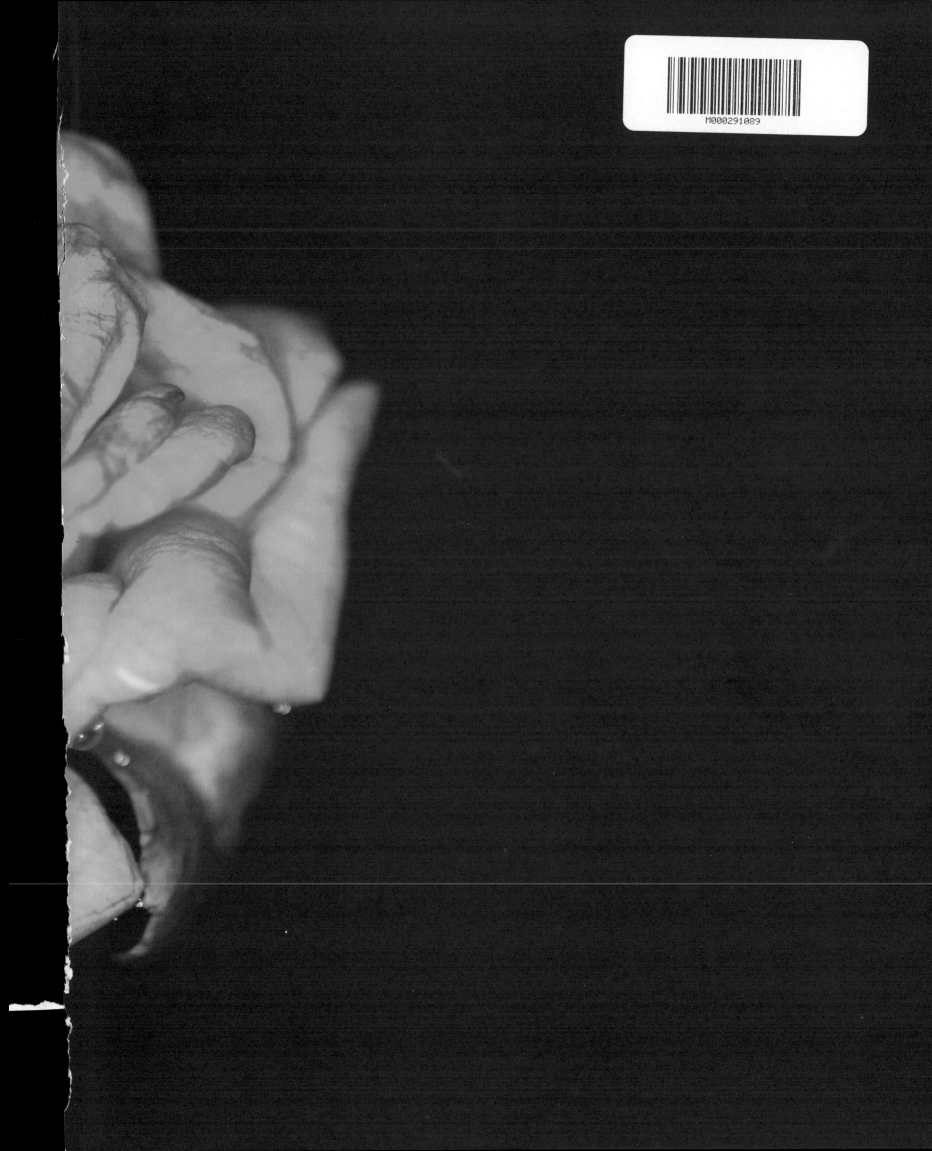

JEFF LEATHAM

JEFF LEATHAM
Visionary Floral Art and Design

RIZZOLI
NEW YORK

New York Paris London Milan

I'LL NEVER FORGET THE FIRST TIME I SAW JEFF'S WORK. It was in the lobby of the George V Four Seasons Hotel in Paris, and the vast scale of his spectacular creations took my breath away. It wasn't just the grandeur of his installations, it was their audacity. He was using form and color in a painterly way I had never seen before in floral design. I remember deep, luscious blooms, the reds so thrillingly beautiful, ranged at impossible angles to dramatic sculptural effect.

Jeff is much more than a floral designer. He is a true artist and a visionary whose impact can be seen around the world. It's no exaggeration to say he has not only revolutionized floral artistry by creating a new style characterized by intense colors, unexpected angles, and grand dimensions, he has also influenced the design world in general.

Foreword

"JEFF HAS NOT ONLY REVOLUTIONIZED FLORAL ARTISTRY BY CREATING A NEW STYLE CHARACTERIZED BY INTENSE COLORS, UNEXPECTED ANGLES AND GRAND DIMENSIONS, HE HAS ALSO INFLUENCED THE DESIGN WORLD IN GENERAL."
NADJA SWAROVSKI

We chose Jeff to create a lighting piece for Swarovski Crystal Palace, our experimental design platform, in which we have collaborated with luminaries such as Zaha Hadid, John Pawson and Ross Lovegrove. We also worked with Jeff one year for the holidays to create a festive display throughout the George V Hotel with chandeliers and magical crystal decorations.

Jeff's exuberant flower arrangements reflect his vivacious personality. His creations possess a unique magic and never fail to bring a feeling of joy to anyone who comes into their orbit. He would have created a profound impact with whatever medium he chose—be it clay, paint, or wood. But Jeff chose the fragile flower. I find his work all the more precious and inspiring for being so ephemeral, and I hope that you, too, are inspired.

NADJA SWAROVSKI

WORKING ON THIS BOOK IT BECAME CLEAR TO ME how much excitement and anticipation I have in sharing my designs. I feel truly blessed, for I think to love what you do is the most important thing in life. But I also believe that with hard work, passion, and perseverance, anything is possible. My hope is that this collection of floral designs not only inspires you, but also encourages you to follow your own design intuition. Walking into the Four Seasons Hotel in Beverly Hills for the first time, I was truly amazed! I saw elegance mixed with great beauty and I knew I wanted to be part of that floral world. I was taught by a patient designer how to think creatively and out of the box. It was all new to me, as I was working with flowers for the first time, with no experience and no formal training. I had only my intuition and passion, which were enough to lead, five years later, to Paris and the role of artistic director of the George V Hotel.

Today, with my team, I create installations and designs that bring moments of magic to the daily lives of the guests. In addition, for my private clients, I design for the exceptional moments in life that call for an extraordinary celebration. Some people say I have a dream job. I couldn't agree more.

My vision is something I hope to share with you through the following pages. We will travel around the world, to some of the most sumptuous spaces and parties on the planet, to some of the most exquisite salons and the coolest events. On the way, we'll get an inside, closeup view of my favorite designs and we'll share the pure thrill of beauty. This book is intended to be a source of inspiration and a reference. It is my sincere wish that it move and energize you to think outside of the box, to find your inner creative confidence, and most of all—to have fun with flowers. Let the journey begin.

Welcome to Jeff's World

"WHAT MAKES HIS FLOWERS SO SEDUCTIVE IS HIS BELIEF THAT THEY ARE AN ARTISTIC MEDIUM . . . HE INFUSES EACH DISPLAY WITH CLEAN LINES, STRONG COLORS, AND A LOT OF UNSPOKEN SEX."
THE NEW YORK TIMES MAGAZINE

JEFF LEATHAM

THEY CALL JEFF LEATHAM
THE ROCK 'N ROLL FLORIST,
but it isn't only because he
wears black. There's something
rebellious in the way he treats
flowers, ripping off their heads,
placing them underwater or
nonchalantly above a vase.
He sees his installations as living
pieces of art. Art that lasts only
in the imagination, or as long as
the evening's event.
Jeff turned the world of flowers
on its head. His designs are bold,
sculptural, often monochromatic
statements of force that use
flowers in the same way an artist
uses color.
There is also a flamboyancy
to the way he breaks stems:
grabbing a bundle of flowers and
snapping their stems over his
knee as if they were firewood.
Jeff exudes confidence and
power, mastering nature and
following its flow—but we're
in no doubt who's boss.

"JEFF'S WORK HAS REDEFINED FLORAL
DESIGN AS WE KNOW IT. I SEE HIS INFLUENCE
IN MY TRAVELS ALL OVER THE WORLD. AND
WHEN I DON'T SEE IT, WHAT'S BEFORE ME
SIMPLY LOOKS OUTDATED.... HIS WORK IS
AN ART FORM, GRAPHIC, ARCHITECTURAL
AND ALWAYS BEAUTIFUL. HIS CHARACTER
SHINES THROUGH HIS WORK AND I HAVE
COME TO LOVE THEM BOTH. HE IS A GREAT
FRIEND AND A GREAT ARTIST."
KYLIE MINOGUE

He introduces a masculinity that does flowers a favor. It's a play of contrasts,
the naked delicacy of a simple flower and the bold direction of its forceful stem.
He revolutionized the way we think about flowers; yet this boldness
is in harmony with the spaces he animates with his creations.
As he says, "It's about creating an experience." Inside these pages we open
the door to the dazzling events Jeff has designed and shine a torch on his
unique vision.

Dense with red roses inside
and out, a twenty-foot corridor
leads into a gala dinner and
concert for a private event
staged in Vienna. First and last
impressions count.

JEFF'S WORLD AS ART DIRECTOR FOR
THE FOUR SEASONS GEORGE V HOTEL is
essentially *parisien*. But his life began in Utah,
in a small town of flat streets enclosed by distant
snow-topped mountains. Summers were spent
in the red-earthed, wide-open spaces near
Bryce Canyon, on his grandfather's ranch.
"I come from a family of cowboys," he says.
The ever-present inspiration of nature and his
"hero" landscape- architect father shaped him
and opened up his visual world.
Fast forward a few years to Paris where Jeff
was modeling and passing a fresh flower market
daily. Here, as he says "my first encounter with
a black rose was the beginning of the love story
of my life."

FROM PARIS AND THE PALACE OF
VERSAILLES TO NEW YORK CITY, from Korea
to the Bahamas, Jeff has worked on some of the
most extravagant and extraordinary parties and
events around the world, creating dramatic floral
designs for more than fifteen years. He enjoys
being hands-on and spontaneous in the way
he creates and his process is fueled by passion
and fun, for him two essential ingredients in
the creative process. These large-scale events
require a great deal of advance preparation and
planning, and these stages are what still excite
Jeff the most.

STRIKING, SCULPTURAL FORMS AND MONOCHROMATIC COLORS play out in a signature style that is architectural in its composition. Jeff sees the vase's shape, measured in the balance of the strong and the delicate, as part of his design, often using a repetition of cylinders, cubes, or spheres. These bring a strong directional anchor against which plucked flower heads can float, or tightly tied leaf-free stems can strike an angled pose. Hydrangeas, calla lilies, orchids, and roses are some of Jeff's favorite flowers.

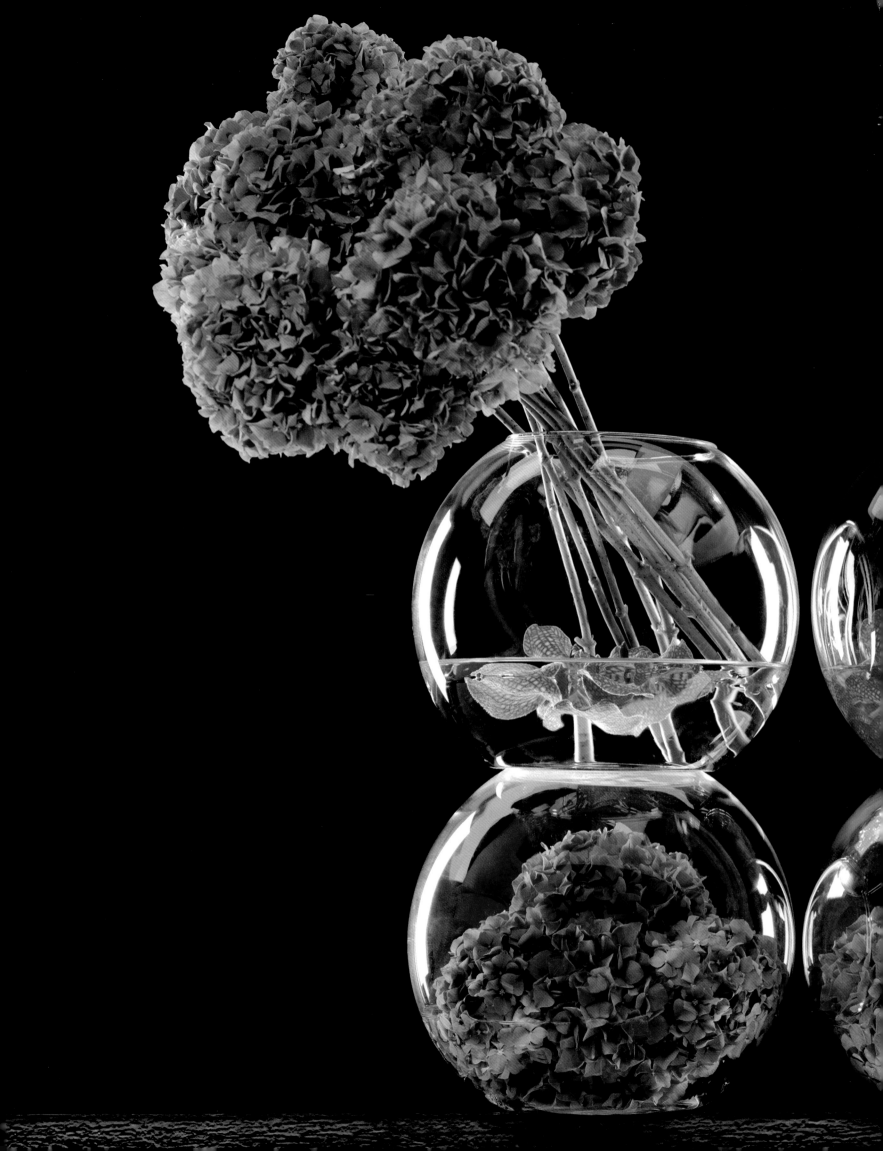

Installations
& Destinations

IT'S ALL ABOUT THE FLOWERS BUT IT'S ALSO ALL ABOUT THE SPACE. Making dramatic design statements in a large area, while still creating a feeling of intimacy, is something Jeff has perfected—and then taken even further.

For him the important word is "harmony" as he seeks out the floral echo of a location. Wherever he is in the world, whatever type of installation he is creating, there is always something organic to his approach. He seeks to complement the venue, adding a gentle softness if the backdrop is graphic, or perhaps bringing a countryside atmosphere onto a New York City rooftop. Finding the right contrast to create a balance is key. A yin to the yang. Jeff strives for harmony, but without sacrificing the boldness of his design. "You can't play it safe when it's about creating that unforgettable experience."

Installations & Destinations

JEFF'S ARTISTRY WITH FLOWERS IS: "THE FLORAL EQUIVALENT TO HAUTE COUTURE-WHEN YOU WATCH A MASTER TAKE A PIECE OF FINE FABRIC AND DRAPE IT, SO THAT IT LOOKS AS THROUGH A CLOUD HAD FALLEN FROM THE SKY TO THE BODY."
SUZY MENKES, FASHION EDITOR
INTERNATIONAL HERALD TRIBUNE

Jeff's rule is to push things as far as they can go, but also to bring intimacy into what are often grandiose and spectacular spaces. To achieve this, he will use candles, smaller-scale arrangements, petals—or delineate smaller, more personal enclaves.

Pink and purple hydrangeas are arranged in monochrome splashes of color for this banquet at the Shilla Hotel Seoul.

PREVIOUS PAGES:
Gladiolus stems echo
the elegant lines of the rooftop
of this Korean temple, adding
a modern twist to a traditional
setting. Fuchsia hydrangeas
anchor the arrangement with
a ring of mokara orchids to add
a contrasting pop of color.

ABOVE AND OPPOSITE:
Jeff's credo is to keep it clean,
simple and chic. These images
taken at the Shilla Hotel Seoul show
how this can be done using baby's
breath and phalaenopsis orchids
to a dramatically elegant effect.
Spherical, clear glass vases bring
movement to the arrangements and
tilt the flower heads, giving them
greater visual impact.

White calla lilies have been used to create a delicate rain-like installation. These are flowers that can last two to three days out of water when hung upside down.

A New York City rooftop
dinner with a woodland
theme, created using bunches
of lily of the valley with
roots and loose bulbs. Tulips
are both strewn over the
arrangement, and placed
on each plate. On the seats
are bags of bulbs, which
guests were given to plant
in their own homes.

FOLLOWING PAGES:
An alternative to the red VIP
rope, this boxwood hedge
is dressed with springeri and
smilax to create a chic private
entrance for this downtown
store opening party.
The surprise on the inside,
like a contrasting jacket
lining, are white stock, peonies,
and roses, which envelop
you and draw you inside.

For the downtown opening of a new store in New York, Jeff used white roses, gardenias, and smilax for a soft touch of romanticism, contrasting with the granite and marble.

For this store, Jeff created
an architectural "cloud" of
hydrangeas, echoing the
graphic black frame with
a modern, lively softness.

FOLLOWING PAGES:
This design using baby's breath
brings a wilder idea to life and
underlines the pure minimalism
of the surroundings. Here, nature
dominates urban sophistication.

The dominating sphere creates a focal point, somehow highlighting the fragility of the stained glass, and also guides the eyes toward the flowers below. Here Jeff used vanda orchids and carnations. The loose petals are scattered, as if the glass had shattered, and the tall black vases echo the grid above.

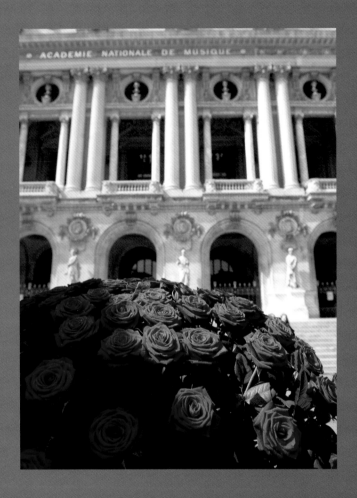

Jeff's designs have adorned some of the most majestic and exquisite historical buildings in the world. The Paris Garnier Opera House, the château of Versailles, and the château Vaux-le-Vicomte are some of the jewels in his crown. In these instances, his first job is to complement the space, to highlight its innate beauty. He works with a strong sense of respect for the space; the flowers are there to beautify the venue, but should always know their place. They are sometimes the star of the show, sometimes the supporting actress.

The principle of less is sometimes more is one that Jeff has always subscribed to. When the mood requires restrained elegance, the flowers set that tone, and when the evening is about riotous fun, exuberance is called for.

In creating these large-scale designs, Jeff tries to anticipate the experience of moving through the space. He creates satellite stories, or individual compositions, that echo the main theme, creating an invisible charge of energy that ties all the elements together. In this way places are turned on, brought to life and reinvigorated, adorned with either simple or sophisticated beauty.

Complementing the location is key. For a setting as opulent as the Opera Garnier in Paris, Jeff enhances the space with a formal, understated approach to the flowers. A black tablecloth shows here how strong lines and perspective can be introduced—and that floral design goes beyond the flowers to consider how they inhabit the space.

For this celebration at the château of Vaux-le-Vicomte in France, red roses underline all the history and tradition while evoking an idea of soldiers on parade. The sprinkling of petals adds a touch of fairy-tale magic.

At the château of Versailles, pink hydrangea heads and horizontal roses bring a blend of modern chic that reinforces the classical beauty of the architecture.

There are few places as majestic as the
château of Versailles, built to reflect the world
dominance and abundant riches of Louis XIV
in the seventeenth century. Being called
upon to create a floral installation here on the
occasion of a state dinner was not only "the
greatest honor of my career" according to Jeff,
but also one of his greatest challenges.
An event on this scale was not new to him,
but on account of the extreme preciosity of the
room, both candles and water were forbidden.
The white calla lilies, orchids, and rose
petals sculpted inside glass are signature
Jeff, and also serve as the perfect echo
to the chandeliers and statues of the *Galerie
des Glaces*.
Jeff approaches a place organically, looking
at the structure and the scale, and then works
to augment and complement it, never to
compete with it.

White rose heads and phalaenopsis orchids create an arrangement that is the epitome of delicate refinement. Hydrangae heads and crystal blush calla lilies in dry vases add a sophisticated dimension that accentuates the pure, raw beauty of the flowers.

Weddings & Ceremonies

WEDDINGS HAVE ALWAYS BEEN ASSOCIATED WITH FLOWERS. Flowers seem to magically amplify the emotion of the day. It's a day of memories, and one that is big in every sense; it's a day to throw caution to the wind and to create a unique experience.

Jeff's work designing these ceremonies that celebrate love is some of his most breathtaking. A wedding is a moment anchored in tradition; it is one of life's most important rendezvous, but it is also a unique, personal, and intimate occasion—often one that has been dreamt of and imagined for years in advance. Used first by the ancient Greeks as a gift from nature, flowers are now inseparable from wedding ceremonies.

Weddings & Ceremonies

"JEFF LEATHAM IS AN ARTIST. HE IS AN INSPIRATION TO THE EVENT DESIGNERS AROUND THE WORLD."
GRACE ORMONDE, EDITOR-IN-CHIEF
WEDDING STYLE MAGAZINE

Jeff's approach to wedding planning is that, above all, you must trust your intuition; it's a day of sharing, but it's also your special day. His own starting point is to understand the bride as he seeks to reflect her personality and bring to life her vision of the day. At the same time, he always delivers a surprise, something so extraordinarily beautiful that it is beyond anyone's dreams. In the words of Tina Turner, "I asked him to do the flowers for my wedding—and he gave me *Ben-Hur*."

PREVIOUS PAGES:
For this Newport, California, wedding on Pelican Hill, Jeff created simple but opulent pastel rose garlands. Swaying in the gentle breeze, they bring a lively, natural elegance to the existing classical structure.

A minimal, modernist take on romantic purity created with peonies, orchids, and hydrangea for a beachside wedding in the Bahamas. Jeff created white hydrangae and rose cuffs to adorn this huppah at the base and the crown. Fragrant orchid garlands hang from the roof.

ABOVE:
An abundance of white-on-white flowers and votive candles brings a warm feeling of intimacy. Tall, transparent vases lend height and elegance, and at the same time, guests can still converse freely across the table. Exquisitely simple bouquets of peonies were carried by the bridesmaids.

OPPOSITE:
White hydrangeas in a box-hedge style demurely echo the ocean outside. Extremely modern and chic.

Roses, hydrangeas, and viburnums are used to create a modern pagoda contrasting with the timeless Korean temples. White diagonally placed calla lilies with loose petals sprinkled on the ground add a touch of grace. The Shilla Hotel, Seoul.

The use of candles and one kind of flower en masse can create drama and impact without sacrificing elegance. As a backdrop, 3000 roses tied onto silk ribbon form a moving "curtain" full of life.

Jeff also believes that the preparations should be fun. He likes to sprinkle a certain fairy dust that makes such a moment deeply memorable; whether it's placing candles where one might expect flowers, or working throughout the night building a wall of 78,000 roses.

One wedding might require several design stories. For the different moments of the day, Jeff creates a specific world, a particular mood for the aisle, the cocktails, the dinner. The aisle is, he says, particularly important because it's the "wow" moment, the first contact the guests will have with the flowers. Done well, it also creates anticipation for what's to come next… Building a seamless event often requires painstaking work backstage. Jeff always visits the location at least once before the process begins. Some of the biggest challenges can come from the limited time there is to install. Whether the venue is a seventeenth-century château in France, the beach, or the George V, the race against the clock is the same.

ABOVE:
With the stems of the
calla lilies tightly wrapped
together, the bouquets are
placed at an angle, creating
contrast with the abundant
bouquets of pink and white
roses and hydrangeas.

OPPOSITE:
At a Korean wedding,
a crown of teak hydrangeas
with cascading white orchid
leis creates a space filled
with zen-like purity.

For an outdoor ceremony, structures can be used to create floral "clouds" and canopies in the sky. But working with nature can achieve beautiful results on a much simpler scale. Existing elements, such as a wall or a large tree, if planned for in advance, can be brought into the design to avoid the need for additional construction.

For every majestic design, there is always a simpler version. Jeff doesn't subscribe to the more is more philosophy, even if his creations are often on a spectacular scale. It's about creating a day more than an installation, and that means everyone should feel happy and comfortable.

One area where less is more is the bouquet. This part of the flower story in Jeff's vision is simple and pared down. Nothing should distract from the bride. The bouquet is an accessory, not the main story. Typically, he will make two: one to keep and one to throw to the wind for the lucky guests. For a contemporary "trend" bouquet, Jeff might use peonies or orchids. For a traditional note, he is a fan of the delicacy and fragrance of lily of the valley, which has been included in the bouquets of princesses throughout history: Audrey Hepburn, Grace Kelly and Kate Middleton.

A white rose canopy with garlands of dangling orchids creates a romantic, seemingly floating, structure for a beach wedding. Carnation petals are used to create a soft and romantic aisle in the sand. What better reason to take off your shoes?

Jeff created hundreds of bouquets of orchids so that the trees would pop with color. These were then tied to the individual branches at this Turks and Caicos Islands wedding—zen-like purity.

Guests were led to the wedding dinner by the green and pink of a walkway clad in "Green Goddess" calla lilies, mint green cymbidium orchids, pink peonies and floating roses.

The scale of the event should never detract from its intimacy. It's your Elizabeth Taylor moment of glamour, but don't forget that you need to feel like you at the same time. For the day to be full of joy, it should ideally be carefree. As Jeff says, "The last thing you want to worry about is your flowers." He maintains it's important that the bride enjoy the wedding at least as much as her guests. His advice is to have good relationships with your wedding planner and florist and to test as much as you can in advance (the bouquet, the food, the walk-through, etc.). The flowers are one of the two big subjects for a bride as she prepares her day; the other is her dress. The aim is to arrive at the day completely reassured about both.

PREVIOUS PAGES:
For this wedding at the Four Seasons Resort The Biltmore
Santa Barbara, Jeff uses full-stem roses to create a
strong sense of perspective with the aisle, as if pointing
to the mountains behind. The elegant simplicity of white
phalaenopsis, white calla lilies and white hydrangeas
arranged at different heights in this tropical setting make
for a design that is chic and modern.

PREVIOUS PAGE:
Upside-down antique-
mauve hydrangeas
create a low floral ceiling
and bring a magical,
wonderland ambiance to
this garden banquet.

ABOVE AND OPPOSITE:
Closely packed white
hydrangea heads wrap
this wedding ceremony
with large floral tendrils.
Rooted vanda orchids
create a canopy of
modern, elegant purity.

Covering existing chandeliers in flowers takes floral decorations to the next level. Here Jeff used 'Voodoo' orange roses, 'Mango' calla lilies and green hydrangeas.

FOLLOWING PAGES:
For the tables, an abundance of vases and flowers will always increase the sense of occasion. Here Jeff has used succulent cactus as an unexpected detail at a low level with the candles.

116

Grand-scale opulence with the chic contrast of black and white, with a splash of red for romance. White spherical hydrangea heads bring an instant hit of glamour against the black cube vases.

For this Saudi royal wedding, the bridal throne is framed with opulent femininity with cascades of pink roses and white hydrangeas.

Wrapping ivy around the pots is a simple idea that can be used to great effect. Here, a Versaille-themed dinner is brought to life with balls of peonies, roses and hydrangeas in a clever play of scale and size.

For an Aspen wedding, Jeff brings the ethereal elegance of the outside to the party inside, using silver birch trunks and baby's breath to anchor each table with a magical tree.

PREVIOUS PAGES:
For the garden ceremony, Jeff created
a wall of yellow tones, to symbolize
the tree of life, new beginnings,
and each day's rising sun.
Jeff wanted to wrap the guests in the
softness of flowers and used hydrangeas,
baby's breath, and roses to do so.

ABOVE AND OPPOSITE:
Romantic, low-hanging garlands of
Naomi red roses create an intimate
atmosphere for the evening meal,
which is illuminated by the romance of
candlelight. Lighting is very important
to Jeff. The way an event evolves
through the day, from a formal event to
something more celebratory or intimate,
is something he designs with lighting,
too. Here, the wall of roses is illuminated
at night to striking effect.

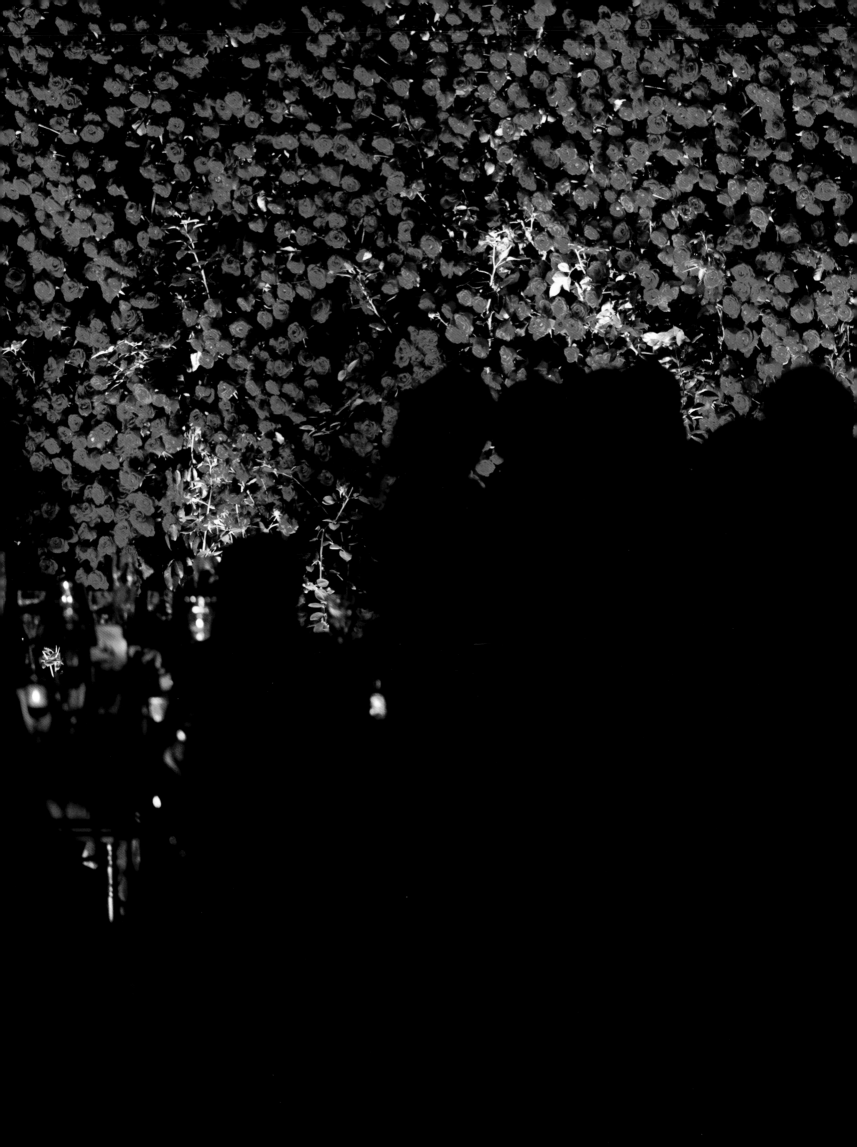

PREVIOUS PAGES:
One of Paris's most beautiful
churches, Saint Germain
l'Auxerrois, was filled with calla
lilies, orchids, and hydrangeas
for this celebrity wedding. Loose
petals along the aisle and low
candles add emotion to the
Gothic grandeur.

FOLLOWING PAGES:
For Eva Longoria's event, the
candelabras were designed to
create a visual connection with
the ballroom at the château of
Vaux-le-Vicomte. Jeff used red
roses as well as calla lilies to
embody passion and love.

For a pop of color Jeff uses hydrangea, fuchsia phalaenopsis orchids and tenga venga rose heads in clear glass bowls for this bridal dinner at the very majestic Baccarat Museum.

Cascading rich reds
create a voluptuous
and opulent sculpture
of red roses, red
peonies, and black
calla lilies for a very
striking posie.

Jeff created a romantic gothic feel for Avril Lavigne's wedding using an abundance of red roses and black candles. The wedding was held in the Château de la Napoule in the South of France.

Table Art

UP CLOSE AND PERSONAL ARE HOW JEFF'S FLOWERS ARE EXPERIENCED when they are the centerpieces of elegant, special-occasion dining. They reveal in some ways a purer iteration of his signature style, as they are independent of the space in which they are found—a riot of poetry on a table that is breathtaking, unforgettable, and chic. Ever sculptural, often monochromatic, they show all the drama of plenty that identifies his style.

Table Art

"I LOOK AT A CENTERPIECE AS A PIECE OF ART, ONE THAT YOU WILL WATCH FOR HOURS."
JEFF LEATHAM

When creating table displays, height is an important consideration. Here, too, "the bad boy of bouquets" breaks the rules, believing that you really can go high. It creates intimacy by inhibiting people from shouting across the table. Early twentieth-century etiquette agrees, as such behavior used to be considered rude. By starting with the vase, the centerpiece can be built with a certain balance and proportion between the flowers and what they stand or float in. Simple spheres, cubes, or cylinders are favorites of Jeff. Stacked, grouped together on a round table, or weaving down the table when there's length, they can add a sensation of movement. Once the vases have structured the design, they just need to be "dressed."

FOLLOWING PAGES:
Closely packed pink roses and carnations with rose petals under glass make for a graphic and bold table design. Jeff never hesitates to deconstruct the flower and to re-create it again differently, as with these transparent petal-filled "table mats."

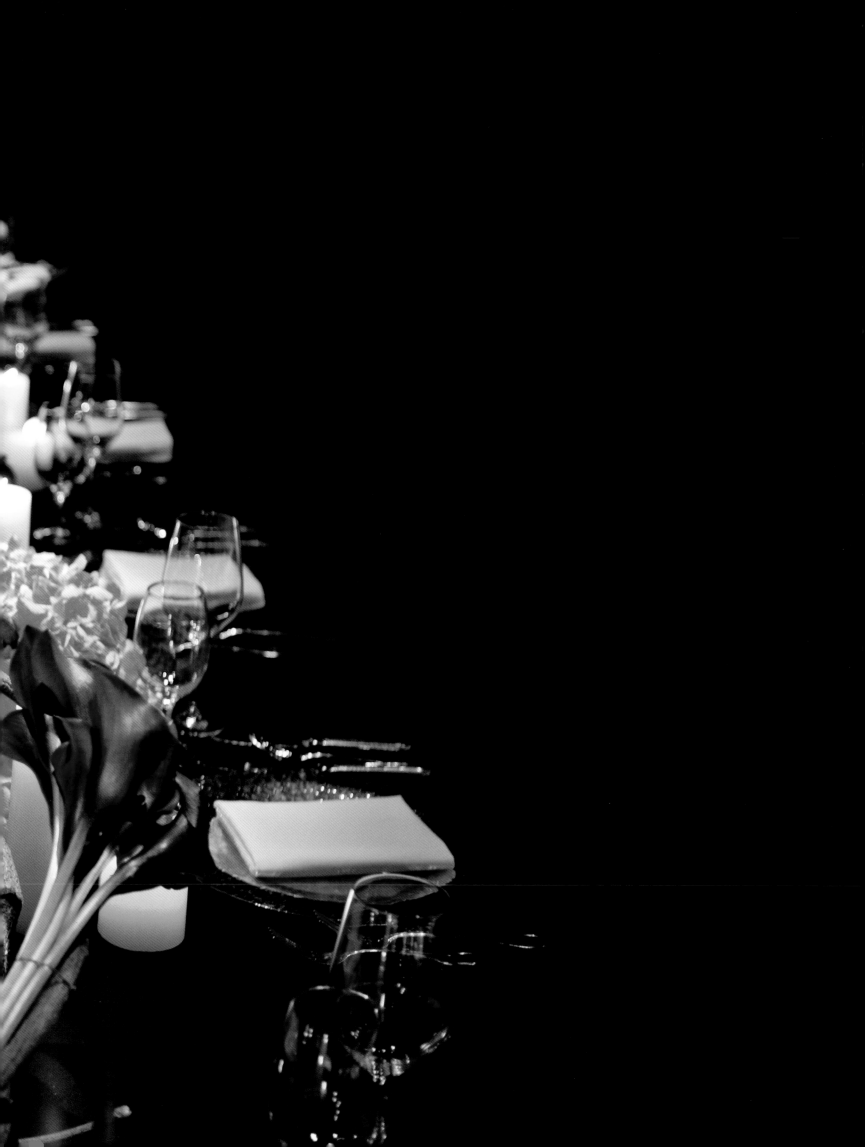

PREVIOUS PAGES:
Hydrangea heads and
calla lilies are woven
down the table through
wisteria branches
and around scattered
candles. Nature is
brought to the table,
simultaneously wild
and sophisticated.

OPPOSITE:
For a touch of daring,
overflowing baby-pink
hydrangea heads form
a tendril-like trail.

Baby pink roses and hydrangeas are used here in two contrasting designs, one sculptural, one traditional. Jeff often creates more than one table design when the room is large to add movement and life.

PREVIOUS PAGES:
Nipped flower heads, when used with clean, simple vases, round or square, create an original display of flowers. Carnations, hydrangeas, orchids, and peonies are all used here without their stems.

LEFT AND RIGHT:
Peonies, paphilopedilum orchids and roses—a truly modern take on romantic dining.

FOLLOWING PAGES:
A romantic tone-on-tone arrangement of roses, hydrangea heads in water, and orchids. Using the same kind of flower inside and outside a vase creates artistry *à la table*.

176

ABOVE:
Props can be used to add interest to a table composition, introducing additional dimensions of shape and color.

OPPOSITE:
Carnations, often discarded in floral design, can look incredibly chic. Here, a tone-on-tone design with a smattering of canda orchids makes for a striking centerpiece.

For Jeff, mirrored vases are a wonderful way to multiply the impact of the centerpiece, creating interesting ricochets of shape and color around the table. Here, clusters of orchids and roses in mauve tones create a patchwork of color.

PREVIOUS PAGES:
Playing with the height of the candles changes the mood. Here, raising the candles increases the formality of the table. Lower candles bring intimacy.

ABOVE AND OPPOSITE:
A floating manzanita branch is an elegant way to drape orchid leis, a design particularly suited to narrower tables where space is at a premium. Another way to bring interest to the table is with a decorative cloth; here, the petal cuts in the cloth add to the sophistication.

OPPOSITE:
This composition tells the story of sensual romance in opulent color saturation. Here, Jeff has tied individual bunches, one flower type in each, and then brought them together in a central podium vase. The flowers lying below, like reflections of the central piece, are reminiscent of a classical still life.

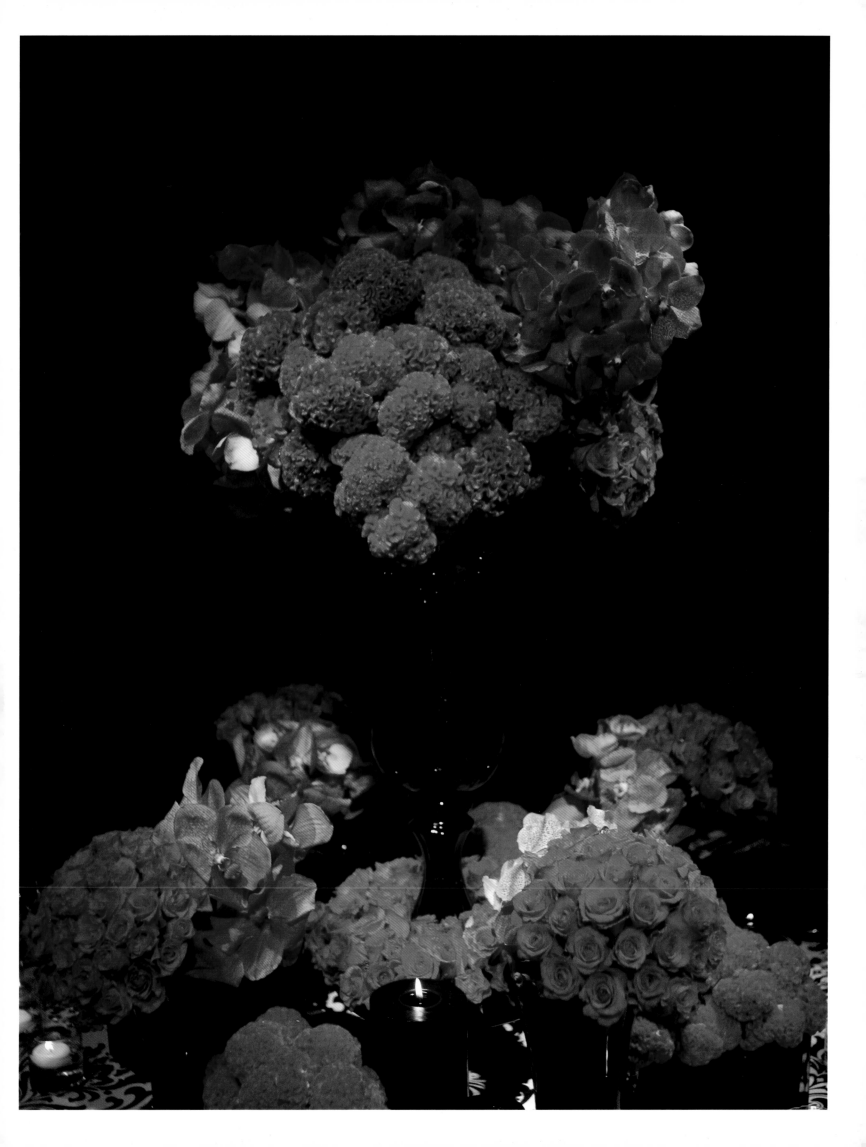

As if growing abundantly from the table, these red roses are in fact held in place with simple Oasis floral foam. Flower heads spread at the base is a clever way to hide any unsightly angles.

OPPOSITE:
This low, meandering stream of petals becomes even more striking with the simple addition of a graphic row of candles.

LEFT:
Breaking his own rules, Jeff haphazardly mixes color in an Indian-inspired riot of dense fuchsia and orange.

OPPOSITE:
The powerful simplicity of an orchid plant magnified behind glass with gracefully draped black calla lilies above.

TOP LEFT:
An original approach
to floral table art, rose
heads are enclosed
under a glass tabletop
that is then dressed
with a mix of votive
and pillar candles.

Very simple flowers, and even grasses, when treated in a clean, simple, and chic way can look sensational.

The color theme is chosen to create an ambiance or to reflect the personality of the host or the event. Preferably the colors should be tone on tone. Similarly, there should never be more than three different varieties of flower. Despite a homogenous approach to color and shape, Jeff likes the design on each table to be slightly different, each echoing the other and making the whole feel more alive. Wrapping the stems tightly, by flower type, before placing them in the vase creates a greater density of color and boosts the power of the blooms. Placing them at a slight angle allows you see them head on when seated. With a 360° approach, this can be repeated around the table. "Not a perfect ball, more like a mountain with hills and valleys," says Jeff, "creating a feeling of movement and mystery." These bombs of beauty are able to transform otherwise unremarkable spaces.

Jeff, sometimes working through the night, is often seen adding the final, decisive finishes, maybe a last touch of petals or candles, just before the first guests arrive. Whether inside a transparent, waterless vase, or directly on the table, the painterly touch is present.

FOLLOWING PAGES:
Wrapping a plain-shaped vase with fabric or ribbon is an additional way to personalize an arrangement and underline the presence of the vase.

ABOVE:
Jeff's tables are often laden with candles, an essential element when creating an intimate dinner.

RIGHT:
Candles, white hydrangea, mirrored vases and slanted calla lilies—pure Leatham.

Sculptural purity with swanlike
grace is created using simple tall
white vases and vanda orchids.
Breaking up the display with
candles at different heights avoids
anything feeling overly stark.

Four
Seasons
Hotel
George V

THE GEORGE V IS ONE OF THE MOST PRESTIGIOUS HOTELS IN THE WORLD. Part of the fabric of Paris's heritage, it's an iconic jewel in the category of *über-luxe* palace hotels. Inside it shimmers. Among the things that really set the hotel apart on today's international stage are the flowers. Jeff's floral designs are its *tour de force*—a statement of beauty that shocks only as much as it inspires, and surprises as much as it reassures. The partnership is a long one, and it's why the flowers have become as much a part of the fabric as the eighteenth-century tapestries. The Four Seasons management came into contact with Jeff's work for the first time in Beverly Hills. They had the designer on a plane a week later, moving him to the City of Light as their artistic director. Both were stepping into the unknown. The French are renowned for their *je ne sais quoi*, the magical ingredient that makes something

Four Seasons
Hotel George V

"HE HAS TRULY REVOLUTIONIZED THE APPROACH TO FLOWERS AND TRANSFORMED IT INTO A REAL ART. HIS FLOWERS HAVE BECOME A TRADEMARK DESIGN FOR OUR COMPANY AND INFLUENCED THE VISION OF FLORAL DESIGN AROUND THE WORLD. JEFF'S FLORAL INSTALLATIONS CREATE AN ENVIRONMENT THAT TOUCHES PEOPLE'S HEARTS AND CREATES BEAUTIFUL MEMORIES."
ISADORE SHARP, FOUNDER AND CHAIRMAN OF FOUR SEASONS HOTELS AND RESORTS

beautiful. It's a question of balance, and achieving that balance has been key to Jeff's success. He decided early on that to strike the right note in the context of lavish opulence, his designs needed to serve as a powerfully fresh antidote to the past—and that the best way to honor the great beauty of the hotel was not to compete with it, but instead to create a contrast with it, bringing in new energy. "It needed to be the complete opposite," says Jeff.

FOLLOWING PAGES:
A symphony of white phalaenopsis orchids that, with the use of repetition and variety, clear glass and mirrored vases, creates a magical moment of wonder in the lobby of the George V.

The commitment of the George V to their floral identity is a story told
in numbers. Jeff and his team of nine florists receive an incredible
15,000 stems every week. Budgets are a closely guarded secret. The fresh
flowers leave Amsterdam in the dead of night three times a week and are
delivered silently to the hotel. The quality of the stems is paramount and the
displays are refreshed around
the clock.

Every four weeks a page is
turned—a new creative concept,
a different cascade of color and
shape, weaves itself through the
hotel. First is the lobby; this is
the wow, the first impression, the
soul of the hotel, its identity.
A moment of awe and wonder
turns into a journey of delight
as you move through the hotel.
Each corner bears a new surprise
and an explosion of decadence
or color that serves to punctuate
every perspective. At night
the atmosphere changes and
candles sculpt the displays with
light, creating a mood as seductive
as it is elegant.
Color is a critical part of the design
process. It is Jeff's starting point.
He can sometimes be found in the lobby watching the space for up to an
hour, deciding on the next ambiance, on the new emotion he wants to evoke.
In this he has carte blanche. Once he has chosen the color or colors, he then
selects the flowers, with his contact in Amsterdam updating him on what
he can source.

OPPOSITE:
Timeless peonies, wrapped tightly
at the stem and positioned at a slight tilt,
reveal their full romantic femininity.

FOLLOWING PAGES:
Pink peonies, deep purple hydrangeas, and rooted
vanda orchids in a contrasting burst of color
in the lobby of the George V. The visible stems
make this arrangement particularly dynamic.

PREVIOUS PAGES:
Giant elephant ear
leaves make for a tropical
paradise and otherworldly
nature, showing the
transforming effect of
flowers.

OPPOSITE:
Vibernum and French tulips,
with carnations sprinkled at
the base. Each installation
is first and foremost about
color, and by creating layers
of color, even simple flowers
with the right props can
create a feeling of *grandeur*.

PREVIOUS PAGES:
The ground floor *galerie* dressed
in an exquisite contrast between
the early eighteenth-century
tapestry and the strikingly modern
and bright orchids.

OPPOSITE:
Cascading vanda orchids
and button roses greet visitors
to the hotel bar. Petals strewn
around the base complete
the sculptural design.

The George V lobby is the perfect stage for Jeff's floral displays. An opera of color greets arriving guests, a modern interpretation of a what an opulent palace feels like today.

PREVIOUS PAGES:
The Louis XIII Salon, filled with classical red roses for formal dining, illustrates how flowers can be used in an understated, simple way to enhance the historical ambiance of a room.

LEFT AND OPPOSITE:
When clients are known to the hotel, the flowers in their rooms are often customized to reflect their preferences and taste. The extreme elegance is reflected in the Fleurology Collection of Waterford Crystal, also designed by Jeff.

FOLLOWING PAGES:
A collaboration between Elie Saab and Jeff Leatham in the Cour de Marbre at the George V Hotel.

Jeff's goal is to create a thrill and to put smiles on peoples faces. He hopes to inspire moments in which one can disconnect from the daily hubbub and dream, if only for a few minutes. His biggest pleasure is creating strong impressions (that famous "wow") that make guests come back for more. In fact, one reason for the frequent design rotation is to surprise returning patrons. For this Jeff receives handwritten notes of thanks from presidents, pop stars, and cultured perfectionists, as well as rock stars and royalty. This evolving, living design is underpinned by some Jeff classics. His love of vanda orchids, hydrangeas, and calla lilies structure many of his creations. For six months of the year Jeff's flowers extend into the open-air courtyard, the Cour de Marbre, bringing the outside that was inside back out into the day. A canopy of raining orchids to compensate for any absence of blue sky was inspired by the techniques for growing, in Amsterdam and Thailand, this exotic, sensual flower. The design, in the words of Jeff, is "a playground of orchid heaven," which once again allows him to transport the guests of the hotel into another world.

Purple hydrangeas, roses, orchids and calla lilies, for an outside, private event in the Cour de Marbre.

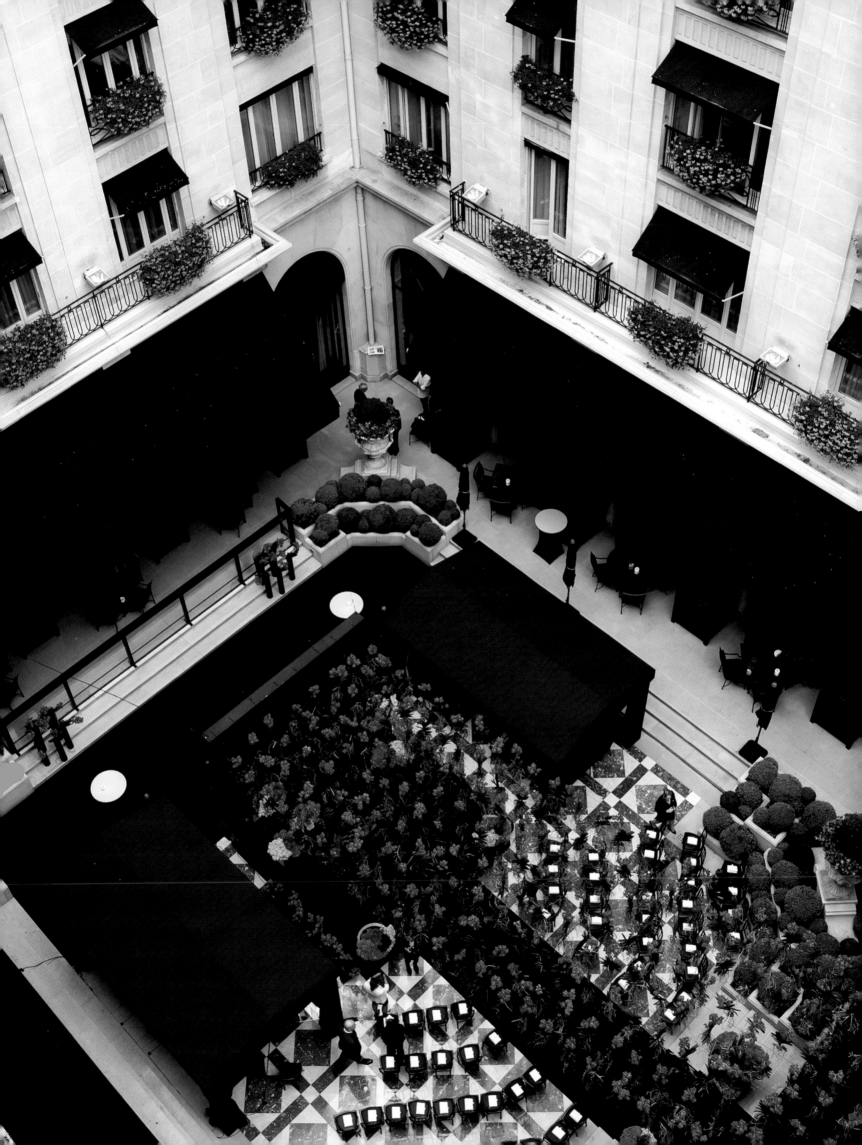

Towering air-born
vanda orchids appear
to be growing sky
high, forming a striking
canopy of vivid pink and
marking the beginning
of the summer season
for the Cour de Marbre.

The view from
the penthouse
suite of the George V,
sculptural romance
in the City of Love.

258

In the words of one of his admirers, Jeff has "the power to make even Christmas sexy." Sexy and modern. At a time of year when the volume gets turned up on visual displays all over the city, expanding his design lexicon beyond the medium of flowers was a natural step. Jeff's passion for contemporary art shines through when designing these majestic installations, expressions of festive magic that are staggeringly chic. Gold-mirrored reindeer standing fifteen feet tall, a courtyard Christmas tree made of eight hundred dangling lightbulbs swaying magically with each gentle breeze, or, collaborating with Swarovski to create walls of sparkling rain, showcasing the talented artists Yves Behar and David Rockwell. Each time, it's a transforming vision that makes adults feel the same kind of excitement they felt as children—and children think they are living inside a fairy tale. For two months of the year Jeff's designs at the Four Seasons George V Hotel are without flowers—but the wow, the magic, and the inspiring beauty is just the same.

FOLLOWING PAGES:
Giant gold-mirrored reindeer are a modern take on Christmas in the lobby of the George V.

PAGES 264–265:
A magical Christmas tree created entirely from lightbulbs.

Acknowledgments

I WOULD LIKE TO THANK THE FOLLOWING DEAR FRIENDS
AND WONDERFUL CLIENTS WHOSE SUPPORT AND KINDNESS,
AND THE OPPORTUNITIES THEY HAVE GIVEN ME, WILL ALWAYS
BE APPRECIATED AND NEVER FORGOTTEN.

Alexander Wang

Alisa Kitamura

Alex Ruiz

Alan Shukovsky

Amanda Hutchinson Watkins

Amanyara Resort - Turks and Caicos

American Foilage

Anco Pure Vanda

Angkhana Chermsirivantana

Arabella Stirling

Avril Lavigne

Admirable Crichton London

Baccarat Cystal

Balenciaga

Belldegrun Family /
Arie and Rebecka,
Ben and Kelly

Bernard and Joan Carl

Billy Evers

Bryan Rafenelli

Carlos Jereissati

Carolyn Cavanough

Catherine Schueler

Chateau Vaux Le Vicomte

Chad Brown

Cher

Christopher Norton

Christian Tortu

Christian Clerc

Christian Felix

Courtney and Tim Presutti

Chateau de la Napoule

Chateau De Versailles

President Bill Clinton
and Secretary of State
Hillary Clinton

Chelsea Clinton and
Marc Mezvinsky

Danny Yezerski

Debbie and Richard Jelinek

Diana Dolan

DIA Art Foundation

Didier Le Calvez

Didier Ubersax

Doug & Lily Band

Dutch Flowers NYC

Eboni Nichols

Edward and Jene Mittler
Brown

Edouard Pailloncy

Elie Saab

Environmental Arts
Turks and Caicos

Erica Moreno

Eva Longoria

Evelyn Herrera

Fischer & Page NYC

Floral Supply Syndicate
Los Angeles

Franzoli Catering Zurich

Frank Ornawski

Fransen Roses

G. Page Wholesale
Flowers NYC

George Heimink

Gebert Family

GM Floral Supply

Grace Ormonde

Givaudan Team NYC

Huma Abedin

Howard and Cindy Rachofsky

Images by Lighting

Irene Boujo

Iguatemi Sao Paulo

Isadore and Rosalie Sharp

Jamali Garden NYC

Jared Rosenquist

Jill Belasco

Jinda Pongpanvadee

Joan Vass

Joann Gregoli

John & Dianne Moores

John Washko

Josh Wood Productions

Katerina Jebb

Kameron Gad

Keith Baptista

Kristen Daniel

Leanne Buckham

Leah Marshall

LM Flower Fashion

McIngvale Family

Marjeth Cummings

Mary Symons

Mathieu Miljavac

Mayesh - Chris Dahlson
& Isabelle Buckley

Martha Stewart

Melissa Torre

Meijer Roses

Michael Pellegrino

Michael Nolan

Michael Venidicto

Mindy Weiss

MoMA NYC

Nadine Jervis

Nadjia Karkar

Nadja Swarovski

Natalie Smith

Nori Bruno

Nick Barraud

Nuage Designs

Oprah Winfrey

Oprea Garnier Paris

Parke Striger

Paul Mather

Paul Tsang Diaz

Qatar Royal Family

Queen Latifah

Raul Lamelas

Roger Thomas

Robert Redford and
Sibylle Szaggars Redford

Revelry Design

Rhonda Graam

Romina Manolio

Rose Marie Bravo

Roxane Debuisson

Samsung

Saskia Havekes

Shawn Gibson

Sherwin Mahone

Sheela Levy

Shilla Hotel

Shilla Hotel Flower Team

Sannie Boers

Suzy Menkes

Terrance Singleton

Taylor Waxman

Todd Shemarya

Van Vliet NYC

VDL Cosmetics Korea

Veronique Abramovitz
Galla Organization

Viceroy Resort Anguilla

Vollering Hydrangea

Walter Grootscholten
Phalaenopsis

Waterford Crystal

Yessid Ortiz

Photographers

THANK YOU FOR KEEPING MY WORK ALIVE LONG AFTER THE "PARTY" IS OVER. I AM EXTREMELY GRATEFUL FOR YOUR TIME, TALENT AND PASSION.

Rebecca Lievre

Florian Kalotay
(www.kalotay.com)

Bob and Dawn Davis
(www.bobanddawndavis.com)

Frederic Lama

Black Alpaga (www.black-alpaga.com)

Studio Cabrelli Paris (www.lescabrelli.com)

Xavier Bejot (www.xavierbejot.com)

Roberto Frankenberg
(www.robertofrankenberg.com)

Elizabeth Messina
(www.elizabethmessina.com)

Ira Lippke Studios
(www.Iralippkestudios.com)

Mark Liddell (www.markliddell.com)

Craig Paulson (www.cpaulson.com)

Simone Van Kempen
(www.simonemartin.com)

Steve Tidball

Diana Bezanski (www.dianabezanski.com)

Chung Film Studio- Seoul Korea

Rami Studio - Seoul Korea

Cinawolf Productions (www.cinewolf.com)

Michael Guez (www.librecommelart.com)

Jeff Leatham (www.jeffleatham.com)

Guillermo Aniel-Quiroga

(www.guillermoaq.com)

Jeff's Special Thanks

I FEEL TRULY BLESSED TO HAVE HAD ALONG MY JOURNEY THE FOLLOWING MENTORS, INFLUENCERS, PARTNERS, INNOVATORS AND VISIONARIES WHO BELIEVED IN ME AND GAVE ME THEIR TIME, SUPPORT, LOVE AND THE OPPORTUNITY TO FOLLOW MY PASSION AND DREAMS.

My beloved, inspiring parents Larry and Janet Leatham

My sister Jennifer who inspires me with her passion for life.

Four Seasons Hotels and Resorts

Four Seasons Hotel George V Paris

HRH Prince Alwaleed Bin Talal Bin Abdulaziz Al Saud

Delphine Graness

Studio Fleur Team FS George V Paris

Jorge Morfin

NYC-based Design Team (you know who you are)

Kylie Minogue

Tina Turner and Erwin Bach

Paige Dixon

Mindy Weiss

Marc Boers

Van Vliet NYC / Elizabeth Lauriello and Sunny Kim

Alexander Wang Team NYC

LM Flower Fashion / Marc Knijnenburg and Raymond Boogaarts

Michel Amann Crystal Group

Jeff's Visionary Book Team

THE BRILLIANT TEAM WHO BELIEVED IN MY VISION FOR THIS VERY PERSONAL PROJECT. THEY HAVE SHARED THEIR TALENTS TO MAKE THIS BEAUTIFUL BOOK AN INSPIRING REALITY.

Charles Miers / Rizzoli

Ellen Nidy / Rizzoli

Ausbert De Arce

Louise Rosen

Vincent Lemaire

Patrice Renard

PHOTO CREDITS

Rebecca Lievre / 4, 8, 22, 178, 268. **Guillermo Aniel – Quiroga** / 229, 230, 240, 242, 243, 244, 252, 262, 264. **Steve Tidball** / 222. **Florian Kalotay** / 16, 134, 135, 136, 138, 139, 140. **Elizabeth Messina** / 104, 105, 108, 110, 111. **Bob and Dawn Davis** / 16, 56, 57, 142, 144, 145, 146, 147, 148, 149. **Frederic Lama - Black Alpaga** / 13, 235, 236, 237, 238, 239, 256, 257, 258, 259. **Studio Cabrelli Paris** / 250, 251, 253, 254. **Xavier Bejot** / 13, 16, 248, 249. **Roberto Frankenberg** / 18, 19, 20, 21. **Ira Lippke Studios** / 68, 70, 72, 73, 74, 75, 76, 78, 80, 94, 95, 96, 100, 101, 102, 103, 130, 131, 182, 183. **Mark Liddell/ Craig Paulson** / 150, 151, 152, 154, 156, 158. **Simone Van Kempen** / 112, 113, 114, 116, 117, 118, 121, 122, 123. **Diana Bezanski** / 38, 39, 40, 41, 42, 43, 216. **Chung Film Studio - The Shilla Hotel Seoul Korea** / 26, 28, 172, 173, 180, 181, 193, 194, 198, 206, 220, 221. **Rami Studio - Seoul Korea** / 24. **Cinawolf Productions** / 10, 11. **Michael Guez** / 124, 125, 126. **Jeff Leatham** / 6, 12, 13, 16, 17, 30, 32, 33, 34, 35, 36, 37, 45, 46, 48, 49, 50, 51, 52, 54, 55, 59, 60, 61, 62, 65, 66, 82, 85, 86, 87, 88, 90, 91, 92, 93, 98, 99, 106, 109, 128, 129, 132, 160, 162, 164, 166, 168, 169, 170, 174, 176, 177, 184, 186, 187, 188, 190, 191, 192, 196, 197, 200, 201, 202, 203, 205, 206, 207, 208, 209, 210, 211, 212, 213, 214, 217, 218, 224, 226, 228, 232, 234. **Book Front cover** / Chung Film Studio Shilla Hotel. **Book Front cover flap** (photo of Jeff) / Rebecca Lievre. **Book Back Cover** / Roberto Frankenberg.

First published in the United States of America in 2014 by

Rizzoli International Publications, Inc.

300 Park Avenue South, New York, NY 10010

www.rizzoliusa.com

© 2014 Rizzoli International Publications, Inc

© 2014 Jeff Leatham

© 2014 foreword by Nadja Swarovski

Texts by Louise Rosen

Designer: Patrice Renard

Design Coordinator: Kaleigh Jankowski

Editor: Ellen Nidy

2014 2015 2016 2017 2018 / 10 9 8 7 6 5 4 3 2 1

ISBN-13: 978-0-8478-4348-0

Library of Congress Control Number: 2014942145

Printed and bound in China

Distributed to the U.S. trade by Random House